Air Force Operational Test and Training Infrastructure

Barriers to Full Implementation

MARK TOUKAN, MATTHEW WALSH, AJAY K. KOCHHAR, EMMI YONEKURA, DAVID SCHULKER

Prepared for the Department of the Air Force
Approved for public release; distribution unlimited

PROJECT AIR FORCE

For more information on this publication, visit **www.rand.org/t/RRA992-1**.

About RAND

The RAND Corporation is a research organization that develops solutions to public policy challenges to help make communities throughout the world safer and more secure, healthier and more prosperous. RAND is nonprofit, nonpartisan, and committed to the public interest. To learn more about RAND, visit www.rand.org.

Research Integrity

Our mission to help improve policy and decisionmaking through research and analysis is enabled through our core values of quality and objectivity and our unwavering commitment to the highest level of integrity and ethical behavior. To help ensure our research and analysis are rigorous, objective, and nonpartisan, we subject our research publications to a robust and exacting quality-assurance process; avoid both the appearance and reality of financial and other conflicts of interest through staff training, project screening, and a policy of mandatory disclosure; and pursue transparency in our research engagements through our commitment to the open publication of our research findings and recommendations, disclosure of the source of funding of published research, and policies to ensure intellectual independence. For more information, visit www.rand.org/about/principles.

RAND's publications do not necessarily reflect the opinions of its research clients and sponsors.

Library of Congress Cataloging-in-Publication Data is available for this publication.

ISBN: 978-1-9774-0884-6

Cover: Ali Stewart/U.S. Air Force.

Limited Print and Electronic Distribution Rights

About This Report

The U.S. Air Force is in the midst of evolving its training infrastructure to respond to expected changes in the conduct of warfare—more specifically, the prospect of operations in contested and denied environments, an increased pace of warfare, and the potential loss of superiority across multiple domains in conflict with near-peer adversaries. The objectives of the Air Force's training program are to deliver readiness by building and sustaining operator skills and to provide information for assessing the readiness of individuals; teams; and, ultimately, of the joint force. Yet, senior U.S. Department of Defense leadership is increasingly concerned that the current readiness assessment system is not providing sufficient insight into the capability of the force to meet future mission requirements—that there is a shortfall in the quality of inputs, and therefore the outputs, of the readiness system. If the Air Force makes appropriate investments, its training infrastructure, which in total is referred to as the *operational test and training infrastructure* (OTTI), could provide much more insight into the readiness of the force for future contingencies. In this report, the authors take stock of the Air Force's current technical capabilities, identify barriers toward implementation of OTTI for readiness assessment, and identify what is required for the Air Force to overcome these barriers.

The research reported here was commissioned by Headquarters Air Force A3T and conducted within the Workforce, Development, and Health Program of RAND Project AIR FORCE as part of a fiscal year 2021 project, "The Use of Operational Training Infrastructure—Live, Virtual, and Constructive Environments in support of a Squadron Commanders Assessments of Unit Readiness Reporting."

RAND Project AIR FORCE

RAND Project AIR FORCE (PAF), a division of the RAND Corporation, is the Department of the Air Force's (DAF's) federally funded research and development center for studies and analyses, supporting both the United States Air Force and the United States Space Force. PAF provides the DAF with independent analyses of policy alternatives affecting the development, employment, combat readiness, and support of current and future air, space, and cyber forces. Research is conducted in four programs: Strategy and Doctrine; Force Modernization and Employment; Resource Management; and Workforce, Development, and Health. The research reported here was prepared under contract FA7014-16-D-1000.

Additional information about PAF is available on our website: www.rand.org/paf/

This report documents work originally shared with the DAF on October 5, 2021. The draft report, issued on July 19, 2021, was reviewed by formal peer reviewers and DAF subject-matter experts.

Acknowledgments

We are grateful for input from subject-matter experts and interviewees from the U.S. Air Force, including officials from Headquarters Air Force Training and Readiness Directorate (A3T); Air Force Agency for Modeling and Simulation; Air Combat Command; Air Mobility Command; Air Force Special Operations Command; Global Strike Command; and Air Force Research Lab.

At the RAND Corporation, we thank Tim Marler, our dedicated reviewer, for helpful feedback and valuable insight and Barbara Bicksler for initial edits and feedback on the structure of the report.

Summary

Issue

Senior leaders on the Joint Staff are becoming increasingly concerned that the readiness assessment system is unable to provide credible answers to whether U.S. air forces can meet the demands of high-end conflict and whether individuals and aircrews have developed the right skills to complete their missions in stressful, complex environments. The sense is that the emphasis has not been on possible future scenarios and that readiness metrics do not provide accurate signals of force deficiencies. This ultimately results in decision priorities that do not align with national strategy.

Approach

This report examines the Air Force's operational test and training infrastructure (OTTI), which is responsible for achieving aircrew readiness, and on the processes for assessing skill development and maintenance. The focus is on OTTI for the combat air forces. The authors describe the technical and other supporting components that constitute OTTI and assess the development frontier for each component. The objective is to offer diverse stakeholders a framework they can use to discern the implications of different training infrastructure investments for assessing skills and monitoring readiness from the individual through joint levels. The authors describe interdependencies across different components of OTTI and implications for coordinating and prioritizing investments in those components.

Key Findings

- The frontier of OTTI development generally falls short of providing a capability to assess skills beyond the individual level.
- Progress in certain areas of OTTI—competency models and data analysis—is limited by a lack of foundational knowledge and a lack of consensus; others are held back by organizational, policy, and technical challenges.
- Beyond fundamental research challenges, the lack of incentives to develop interoperable and standardized training systems and the parochial focus of training investments within the DAF and across DoD hinder progress along multiple dimensions of OTTI.
- Progress along different dimensions of OTTI is highly interdependent; for example, the development of data storage and analytic capabilities for assessing readiness in highly complex environments depends on identification of competency models and associated performance criteria.
- More-authentic training at the individual level alone is unlikely to provide an accurate capability to assess collective and joint readiness; more research is required to identify

competencies, metrics, and data analytic capabilities to assess performance and readiness at higher levels of aggregation.

Table S.1. Barriers Toward Advancing the Horizons of Operational Test and Training Infrastructure Technologies

Category	Initial Skills	Individual Skills	Collective Skills	Joint Skills
Competency model	Perceptual and motor skills	Task behaviors	Team behaviors	Interteam behaviors
Infrastructure	Trainer	Simulator	Networked simulators	Distributed joint mission training
Simulation capabilities	Low environment complexity	High environment complexity	Red air models	Diverse Blue and Red system models
Data capture and storage	Few performance variables	Many performance variables	Scenario and Red air	Non–Air Force systems
Data analysis	Performance relative to standard	TTP selection	Team behavior (communication)	Interteam behaviors

NOTES: Green shading indicates that the limitations are related to technical, organization, or policy issues. Blue shading indicates that the limitations are more fundamental (lack of consensus, foundational knowledge). The black line indicates current horizon boundaries.

Contents

Figure and Tables

Figure

Tables

1. Introduction

Senior leaders on the Joint Staff are becoming increasingly concerned that the readiness assessment system is unable to provide credible answers to whether the U.S. Air Force can meet the demands of high-end conflict and whether aircrews have developed the right skills to complete their missions in stressful, complex environments. The Air Force Chief of Staff, Gen Charles Q. Brown, and the U.S. Marine Corps Commandant, Gen David H. Berger, recently wrote that the focus of readiness is, instead, "inappropriately weighted in favor of what is available to fight a narrow range of scenarios today with what we currently have on hand."[1] Current readiness metrics thus do not provide accurate signals of force deficiencies, which, ultimately, results in decision priorities that do not align with national strategy.

The Chairman of the Joint Chiefs of Staff evaluates whether U.S. forces are capable of accomplishing national security objectives through a self-assessment program known as the Chairman's Readiness System (CRS).[2] The CRS collects information on whether units have adequate resources by assessing whether units meet benchmarks for amounts of trained personnel and functioning equipment. Many other factors also contribute to readiness; therefore, the system also collects subjective information directly from unit commanders on whether their units can accomplish all mission tasks. Policymakers commonly refer to these two dimensions as *resource readiness* and *capability readiness*, respectively.

The objective of the Air Force's training program is to deliver readiness by building and sustaining operator skills. Assessing resource readiness in the training domain involves tracking inputs, such as the frequency, recency, and types of training events that an individual has completed against predetermined readiness standards. Then, commanders can identify further training deficiencies in capability readiness reporting if the training standards do not tell the whole story.

The senior leaders are not questioning the readiness concepts inherent in the CRS; rather, they are highlighting shortfalls in the quality of the inputs, and therefore also the outputs, of the readiness system. A key challenge that limits the utility of both resource and capability readiness inputs is that information is lacking on how to connect inputs to the most important outcomes— the capability of Air Force pilots to conduct the full spectrum of operations to defeat all enemies.

Historically, the best way to generate empirical data on force capabilities is through major exercises. These exercises are infrequent and limited in scope, because they require immense

---

[1] Charles Q. Brown, Jr., and David H. Berger, "Redefine Readiness or Lose," War on the Rocks website, March 15, 2021.

[2] Chairman of the Joint Chiefs of Staff Guide 3401D, *CJCS Guide to the Chairman's Readiness System*, Washington, D.C., November 15, 2010.

resources (often the same resources that would otherwise generate more readiness).[3] However, there might be other ways to create training experiences that produce better feedback on how capabilities perform. Better-quality simulated training and data-collection capabilities, for instance, create opportunities to answer the call for better readiness information by going beyond the current system of tracking the frequency and types of training events. Moreover, improving readiness information has other less-obvious limiting factors that policymakers must also consider.

The objective of this report is to identify barriers to developing a training infrastructure that can put the Air Force on a trajectory toward improved readiness assessment.[4] Our focus is specifically on the combat air forces, although the report has broader implications for operational test and training infrastructure (OTTI) across the Department of the Air Force.[5] Our approach can also be used to identify OTTI development priorities over time. A more mature OTTI will help the U.S. Department of Defense (DoD) and Air Force senior leaders assess whether the Air Force can meet the demands of high-end conflict and whether pilots have developed the right skills to complete their missions in complex joint environments.[6] A companion report will focus on our findings and provide more-specific recommendations for how to use OTTI to improve readiness assessments.

OTTI consists of the technical infrastructure that enables operational training: facilities, simulators, networks, threat models, and data capture and analysis tools.[7] In this report, we take stock of these technical capabilities and identify what is required for the Air Force to leverage the information OTTI provides for assessing readiness.

Designing and prioritizing investments in OTTI is a cooperative effort spanning multiple organizations and functions. We developed this framework as a reference so that diverse stakeholders can easily discern the implications of different investments for readiness metrics as they plan. We expect the framework to be a useful tool for coordinating efforts among (1) personnel in Headquarters Air Force and major command (MAJCOM) readiness directorates; (2) simulation technology experts at the Air Force Agency for Modeling and Simulation or the

---

[3] Richard K. Betts, *Military Readiness: Concepts, Choices, Consequences*, Washington, D.C.: Brookings Institution, 1995.

[4] Note that the approach in this report does not allow a rigorous cost-benefit analysis of different aspects of OTTI.

[5] Although this report focuses on training, OTTI includes assets that support testing of weapon systems, as well as warfighter readiness training. This reflects the current Air Force acquisition, testing, and training concept, which considers all three aspects together to plan for maintaining readiness earlier in the acquisition process.

[6] The focus of this report is on U.S. Air Force training. References to joint training and skills refer to the Air Force's contribution to joint readiness and not joint training events and capabilities per se. For more on differences between Air Force and joint training events Air Force Instruction (AFI) 10-204, *Air Force Service Exercise Program and Support to Joint and National Exercise Program*, Washington, D.C.: U.S. Department of the Air Force, April 12, 2019.

[7] AFI 16-1007, *Management of Air Force Operational Training Systems*," Washington, D.C.: U.S. Department of the Air Force, October 1, 2019.

Air Force Life Cycle Management Center (Architecture and Integration Directorate); and (3) research organizations, such as the 711th Human Performance Wing at Air Force Research Laboratory.

In Chapter 2, we discuss the history of training and advancements in technology. In Chapter 3, we describe current training and readiness processes in the OTTI context. In Chapter 4, we offer a high-level view of the various elements of OTTI stand, where their limitations and barriers to progress are, and how the elements might be integrated to revolutionize pilot training and readiness assessment. In Chapter 5, we summarize our findings and offer concluding thoughts.

2. Recent Advances in Air Force Training Infrastructure

The enhancements to OTTI that are underway as of late 2021 are part of a DoD-wide effort to improve training capabilities so that they better approximate operations in a combat environment, which in turn, creates more-realistic feedback on how forces will perform in such environments. In particular, the U.S. Air Force is developing the Common Synthetic Training Environment to provide a platform-agnostic, distributed synthetic training environment to address a range of OTTI gaps.

Previous training revolutions have often achieved great success. The military has developed technologies and training processes to deliver more-authentic and larger-scale training experiences to prepare forces for future conflicts. This chapter presents a review of previous training revolutions to provide important context for the goals and challenges associated with current efforts to improve the training system and build readiness for high-end conflict.

A Brief History of Department of Defense Training Paradigms

The first training revolution was driven by poor air-to-air combat performance in Vietnam and the growing conventional threat of the Warsaw Pact, the materiel capabilities of which were demonstrated in the 1973 Yom Kippur War.[1] In the wake of these events, the U.S. Navy, Army, and Air Force all instituted more-realistic training to better replicate the stress and intensity of real-world scenarios. This revolution involved larger-scale training events and more realism. After-action reviews, in which feedback from training events informed the content of subsequent training, were a central aspect of the Army National Training Center's contribution to this revolution.[2] Both the Navy (TOPGUN) and the Air Force (Red Flag) exercises instituted training that developed pilot skills against more credible threats and in more-authentic environments.

Within the Air Force, a second training revolution began in the late 1990s that centered on distributed training (distributed mission operations [DMO]). This revolution was aimed at providing training that (1) was dynamic, in that it responded to changes to technology and the security environment; (2) addressed the capabilities of potential adversaries; and (3) shaped a force to fight alongside other military, nonmilitary, joint, and coalition actors.[3]

---

[1] Thomas C. Greenwood, Terry Heuring, and Alec Wahlman, "The 'Next Training Revolution': Readying the Joint Force for Great Power Competition and Conflict," *Joint Force Quarterly*, Vol. 100, 1st Quarter 2021, pp. 26–34.

[2] Anne W. Chapman, *The Army's Training Revolution, 1973–1990: An Overview*, Fort Monroe, Va.: U.S. Army Training and Doctrine Command, 1991.

[3] Robert Chapman and Charles Colegrove, "Transforming Operational Training in the Combat Air Forces," *Military Psychology*, Vol. 25, No. 3, 2013.

Distributed training capabilities were intended to enable the development of individual, collective, and joint skills. The training capabilities that had been developed in the first revolution were aimed at meeting the larger-force scenarios for which the AirLand Battle doctrine was developed.[4] The second revolution has unfolded in the context of combat experiences in the Middle East after September 11, 2001, in which small-unit, tactical training has grown more important.[5]

DMO promises not only better approximation of live training with real equipment on real ranges but also training capabilities that live environments simply cannot provide because of safety concerns, security concerns, or the costs of organizing large events with many participants.

In principle, the networked nature of distributed training permits the training of many participants, with each individual or group of individuals potentially exercising a different function. The ability to assess readiness across a range of levels—from the individual to joint levels—and across joint functions may seem just a short step away, given the existing capability to network training devices in simulated environments.

However, the technical and supporting capabilities needed to achieve this vision are not in place, even while the second training revolution has built upon the first in providing more-authentic training experiences. As the chapters that follow will detail, the Air Force still struggles in many respects to provide the dynamic, capabilities-based, and networked training that the DoD envisioned for U.S. forces over 20 years ago.

Current Developments in the Department of the Air Force's Approach to Aircrew Training and Readiness

Even though the second revolution of DMO remains incomplete, the Air Force is simultaneously pursuing two further paradigm shifts. First, building on the unrealized aims of the previous training revolution, the Air Force wants to enable training *across domains and services* in *highly complex environments*. Second, current efforts to improve training and readiness are focused on developing aircrew *proficiency*.

Just as the training adaptations of the 1970s and 1980s shaped a force to fight according to the AirLand Battle doctrine, the emerging doctrine of Joint All-Domain Operations (JADO) outlines the factors motivating the shifts toward assessing proficiency in complex, highly

---

[4] AirLand Battle doctrine was the Army's operating concept that emerged in the 1980s, placing a focus on the operational level of warfare and integration of the air and land domains to counter the Soviet threat in Western Europe.

[5] Robert H. Scales, "The Second Learning Revolution," *Military Review*, Vol. 86, No. 1, 2006.

networked training environments.[6] Unlike the previous training revolutions, which were motivated by battlefield results, the current paradigm shifts are driven by expectations of how the conduct of warfare will change in the near future, focusing on near-peer threats, the prospect of operations in contested and denied environments, increased pace of warfare, and the potential loss of superiority across multiple domains in a future high-end conflict.[7]

The second paradigm shift, toward a proficiency-based training and readiness system, is desirable in its own right, putting aside the strategic imperatives that are motivating the emerging JADO doctrine. The existing readiness system is based on currency: Aircrew are considered ready if a minimum number of sorties and training experiences have been accomplished within certain time frames.[8] A currency-based approach to readiness implicitly assumes that, for instance, a certain number of sorties is a reliable predictor of readiness for a given aircrew role. However, target counts may be either too little *or* too much; worse still, in some instances, a given type of sortie may not increase proficiency in the desired skill set.[9]

A proficiency-based training system aims to target the performance levels of training audiences by tracking a set of performance metrics. In a proficiency-based system, training could be targeted toward specific sets of high-priority skills. Fine-grained, objective performance measurement also permits the reallocation of resources by identifying which sorts of training capabilities produced the greatest increases in proficiency. A proficiency-based system thus provides tools to answer such questions as the following: "How much does the air-to-air missile hit rate fall if I cut flying hours in half?"[10] Answering questions like this is substantially more difficult at the unit and joint levels, but some of the same building blocks that enable individual assessment are also ingredients in the more complicated enterprise of assessing readiness at higher levels.

The two paradigm shifts are distinct but related in two important respects. First, the inability to systematically collect objective data hinders the ability to implement a proficiency-based training system; it would be too labor intensive to rely on subject-matter expert (SME) judgments to evaluate proficiency at scale. However, the simulators and simulation capabilities needed for networked, mission-integrated training can potentially provide such objective data.

---

[6] JADO comprises "air, land, maritime, cyberspace, and space domains, plus [the electromagnetic spectrum]. Actions by the Joint Force in all domains that are integrated in planning and synchronized in execution, at speed and scale needed to gain advantage and accomplish the mission" (Air Force Doctrine Note 1-20, "USAF Role in Joint All-Domain Operations," March 5, 2020).

[7] James B. Hecker, "Joint All-Domain Operations," *Air & Space Power Journal,* Summer 2021.

[8] Paul E. Carpenter, *Getting More Bang for the Buck: Incentivizing Aircrew Continuation Training*, Maxwell Air Force Base, Ala.: Air University, December 2016.

[9] The Air Education and Training Command's Pilot Training Next program for undergraduate pilot training is using elements of proficiency-based training.

[10] Todd Harrison, "How Many Flying Hours Does It Take to Kill a Terrorist?" Defense One website, November 17, 2014.

Second, a more-fully realized networked training capability across the joint force would help enable the objective assessment of readiness at different levels—individual, unit, and joint—as envisioned by the second training revolution.

Together, the paradigm shift toward proficiency-based training and a more complete realization of networked, mission-integrated training in the Air Force hold great potential to build readiness and inform assessment. However, as the sections below detail, challenges remain in understanding how to leverage capabilities in these two areas to create the most meaningful information on readiness that reflects how combat power is actually employed.

3. Current Training and Readiness Processes and Where the Operational Test and Training Infrastructure Fits

Although OTTI is used throughout the entirety of the U.S. Air Force training pipeline, from initial skills training through continuation training, continuation training is the primary area in which the Air Force assesses the types of missions that individuals and aircrews are ready to execute. To that end, we detail how OTTI fits within current training processes that produce qualified individuals and aircrews at the squadron level and informs readiness assessment.

Individuals and aircrews attain and maintain various levels of status related to mission-readiness through training that they receive once assigned to a unit.[1] The tasks that units are required to train for are determined by the designated operational tasks that units must be able to execute in combat.[2] Units draw from Air Force and joint task lists and publications in the AFI 11-2 series to identify the essential tasks for which units have training and reporting responsibilities.[3]

SMEs use these documents, along with the output of tactics conferences and training review boards, to create Ready Aircrew Program (RAP) tasking memorandums (RTMs), a central document that the Air Force uses to specify requirements for maintaining certain mission-readiness statuses for different weapon systems. RTMs specify the numbers and types of sorties (live or virtual, day or night, and so on) that aircrews are required to experience to maintain different qualifications over set periods.

Combat and weapon system officers and unit training offices create training plans for aircrews by drawing on these RTMs. Aircrew qualification status is determined by an individual's completion of sorties requirements set out in the relevant RTM.[4] These sortie completion counts are a primary input into the readiness reporting system and the main way that OTTI factors into reporting on aircrew training readiness. The outputs of this process inform the training readiness, or T-level, that is reported in the official DoD readiness reporting system (the

---

[1] Aircrew members attain and maintain Basic Mission Capable, Combat Mission Ready, and other specialized statuses at the unit level.

[2] Individual units have a Designed Operational Capability statement that specifies the tasks that the units must be able to execute in combat.

[3] For an overview of the Air Force's training program, see AFI 11-200, *Aircrew Training, Standardization/ Evaluation, and General Operations Structure*, September 21, 2018. The AFI 11-2 Volume 1 series consists of mission design series (MDS)–specific volumes that address personnel training, qualification, and certification on that MDS.

[4] In addition to continuation training, pilots must complete mission qualification training on arriving at a new unit and must complete flight lead and instructor pilot upgrade training to further develop their skills. These activities are not separately tracked in readiness reporting yet contribute to a unit's ability to deliver and develop combat power.

Defense Readiness and Reporting System–Strategic. A unit's T-level indicates the degree to which a unit's aircrews have completed the required training; most commonly, this is dictated by the unit's RTM.[5]

Commanders also report on their subjective judgment of a unit's readiness to execute mission-essential tasks, reported in the readiness system as a unit's capability readiness. Although aircrew performance in training events that make use of OTTI influence a commander's judgment of the unit's capability readiness (reported as "yes," "qualified yes," or "no"), data from OTTI are not captured in any systematic way across the Air Force to inform this assessment and do not directly factor into readiness reporting. However, individual components of OTTI could be used to improve the training and readiness system described here.

---

[5] AFI 10-201, *Force Readiness Reporting*, Washington, D.C.: U.S. Department of the Air Force, December 22, 2020.

4. The Operational Test and Training Infrastructure Landscape

The OTTI consists of groups of technical capabilities that have been developed at different times and by different communities. For example, the first flight simulator was patented in 1930 and allowed pilots to practice controlling the attitude of a fuselage using a mechanical stick and rudder.[1] The evolution of flight simulators in the time since has followed advances in electronics, digital computers, motion systems, and visual systems and has been led by the defense industrial base and commercial aviation.[2] State-of-the-art simulators currently have a wide field of view, motion along multiple axes for larger multiengine aircraft training devices (such as mobility; bomber; and command, control, intelligence, surveillance, and reconnaissance aircraft), sound, haptic cues in some lower-level devices, and concurrent mission systems to deliver high-fidelity training across a wide range of scenarios.

The first system for automated performance assessment, the Performance Evaluation Tracking System (PETS), did not emerge until the start of the 21st century, although it has not been widely adopted in operational training.[3] PETS was enabled by advances in networked simulators, data capture, and psychological measurement theory that specified the knowledge and abilities that pilots need and the sources of evidence to demonstrate mastery. The development of PETS has been led by the U.S. Air Force's science and technology enterprise.

The next generation of pilot training capabilities will build on research and development activities from across industry, academia, and the military. Many of these activities were not initiated with pilot training in mind yet have potential utility for this purpose. For example, the entertainment industry has created visual and haptic interfaces that can increase simulator fidelity, and academia and industry have developed artificial intelligence methods that can help simulate adversary behaviors.[4]

The OTTI must integrate multiple technologies into a cohesive architecture. Figure 1 gives a high-level view of how the various elements of OTTI may be integrated to revolutionize pilot training along with readiness assessment. Figure 1 depicts the following elements:

---

[1] Edwin A. Link, Jr., Combination Training Device for Student Aviators and Entertainment Apparatus, U.S. Patent 1,825,462, filed March 12, 1930.

[2] Ray L. Page, "Brief History of Flight Simulation," presented at SimTecT 2000, Sydney, 2000.

[3] Brian T. Schreiber, Eric Watz, Antoinette M. Portrey, and Winston Bennett, Jr., "Development of a Distributed Mission Training Automated Performance Tracking System," in *Proceedings of the Behavioral Representations in Modeling and Simulation (BRIMS) Conference*, Scottsdale, Ariz., 2003.

[4] Patrick Tucker, "An AI Just Beat a Human F-16 Pilot in a Dogfight—Again," Defense One website, August 20, 2020.

**Figure 1. How Operational Test and Training Infrastructure
Technologies Fit Within an Adaptive Training System**

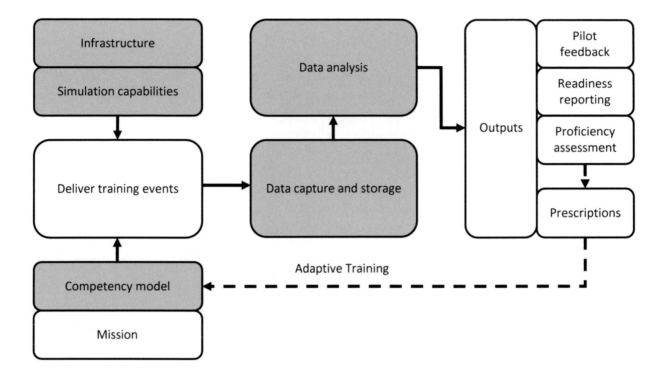

NOTE: OTTI technologies are shaded in blue.

- The **mission** is what an individual or unit is responsible for. This is described in a unit's Designed Operational Capability statement, core mission-essential tasks, and RTMs.
- The **competency model** consists of the underlying individual, collective, and joint competencies needed to complete missions. Mission Essential Competencies (MECs) are one example of a competency model.[5] Elements of the competency model should relate to the mission.
- The **infrastructure** consists of the simulators, ranges, facilities, instrumentation, networks, protocols, and other elements needed to deliver simulation-based training.
- **Simulation capabilities** offer the ability to embed models of friendly systems, threat systems, and environment conditions in live, virtual, and constructive settings.[6] Simulation capabilities depend, in part, on infrastructure.
- The mission, model, infrastructure, and capabilities feed into **delivery of training events**. Infrastructure and simulation capabilities determine the set of events that a training

---

[5] Brian T. Schreiber, "Transforming Training: A Perspective on the Need for and Payoffs from Common Standards," *Military Psychology*, Vol. 25, No. 3, 2017.

[6] Live training events involve real aircrews operating real platforms; virtual events involve people operating virtual equipment (e.g., a simulator); constructive events involve simulated people operating simulated equipment (e.g., computer-controlled adversary forces injected into a simulated environment).

manager can possibly deliver, whereas competency models help the training manager prioritize which events to actually deliver to allow pilots to meet mission requirements.

- Data are generated during the delivery of training events. **Data capture** involves recording information about scenarios, conditions, mission outcomes, live and simulated flight data, objective and subjective performance measures, and other metadata to permit analysis of training events.[7] **Data storage** involves archiving events and data in a persistent database to support debriefing, analysis, and assessment of proficiency and readiness.

- Once captured and stored, data can be analyzed. **Data analysis** involves extracting meaning from training data. Basic analysis focuses on training completion, such as the number of training events pilots accomplish monthly. Analysis can also focus on task or mission outcomes, such as kill ratios. Finally, analysis can produce new information, such as estimates of pilot proficiency based on performance measures.

- Data analysis produces meaningful **outputs** that can be used for assessment and training. Instructor pilots can play back events to provide *pilot feedback* during debriefings. Count-based metrics are used to determine whether pilots are Combat Mission Ready and to report unit training readiness in the *readiness reporting* system (T-Level).[8] Outputs are used to assess *proficiency* and, ultimately, *prescribe* how future training events can develop competencies.

A major element of proficiency-based training is adapting the training to an individual's needs (the dotted arrow in Figure 1). Objective and subjective measures captured in future OTTI may inform performance-based proficiency estimates, which are necessary to deliver adaptive training, and, ultimately, readiness metrics that more accurately address individual and unit ability to perform in high-complexity environments alongside joint forces.[9] Furthermore, performance data from the future OTTI may also be used to select scenarios to adaptively test individuals' readiness.

The Air Force vision for OTTI is motivated by the needs for effective training and for performance-based readiness assessment. To meet these needs, the Air Force must decide which technologies to invest in and when. These decisions can be informed by understanding the capabilities that exist today and the capabilities that would be required to extend training and assessment from individual skills to collective and joint skills.

We identified technology categories by reviewing OTTI technical documentation and academic literature and by conducting interviews with SMEs. Table 1 shows five general categories of OTTI technologies (rows) along with four skill categories (columns). The skill

---

[7] Jeffrey M. Beaubien, Michael Tolland, and Jared Freeman, "Performance Measurement Applications and Associated Data Requirements for LVC Training," presented at the Interservice/Industry Training, Simulation, and Education Conference, November–December 2020.

[8] AFI 10-201, 2017.

[9] The discussion in this chapter notes some types of data (see "Data Capture") and predictive analytic capabilities (see "Data Analysis") that would enable this. However, it is beyond the scope of the report to recommend the adoption of specific measures or analytic capabilities. Indeed, we argue below that data types and analytical tools should be tightly linked to advancements on competency models.

categories represent isolated initial skills, the effective combination of initial skills that gives rise to individual skill, collective skills, and joint skills. The technology categories represent the subset of OTTI technologies prior to the output stage of the adaptive training process depicted in Figure 1.[10]

Table 1. Categories of Technologies Comprising the Operational Test and Training Infrastructure

Category	Initial Skills	Individual Skills	Collective Skills	Joint Skills
Competency model	Perceptual and motor skills	Task behaviors	Team behaviors	Interteam behaviors
Infrastructure	Trainer	Simulator	Networked simulators	Distributed joint mission training
Simulation capabilities	Low environment complexity	High environment complexity	Red air models	Diverse Blue and Red system models
Data capture and storage	Few performance variables	Many performance variables	Scenario and Red air	Non–Air Force systems
Data analysis	Performance relative to standard	Selection of tactics, techniques, and procedures (TTP)	Team behavior (communication)	Interteam behaviors

The capabilities across the five categories shown in Table 1 collectively determine the set of skills that can be assessed objectively and in an automated manner. For example, to train and automatically assess performance of basic pilot skills, the Air Force must, at a minimum, achieve the set of capabilities contained in the second column, "Initial Skills." Although it may be beneficial to advance the capability level in any one OTTI technology category, training and automated performance assessment are limited by the least mature technology across the complete set of categories.

The next sections describe each row of Table 1 in detail and discuss the current frontier and the horizon for future development. The frontier is meant to demarcate the current state of the art: a set of capabilities that have been demonstrated and are in regular use in operationally representative training environments.[11] For example, we identify the skill level up to which the Air Force's current simulation capabilities allow it to regularly train in such environments. We set frontiers based on our literature review and SME interviews. For the literature review, we consulted academic and government literature on the use of modeling and simulation, training, and readiness. Interviews included a range of personnel involved in training and readiness from MAJCOMs and from industry.

---

[10] In this discussion, we combine data storage and data capture and treat missions as exogenous, or external to the process of adaptive training.

[11] The challenges for collective skills are exacerbated for joint skills because the latter are even more likely to draw on geographically dispersed units with different weapon systems and from different services.

The discussion also identifies barriers to advancing individual components to address the skills that aircrews require to perform in collective and, ultimately, joint contexts. Although we do not make recommendations to move forward on any particular component of OTTI, we identify what would be required to do so. The discussion that follows is not an exhaustive, in-depth review of individual components of OTTI but instead summarizes the current state of each component and provides background for a framework for identifying barriers to implementing OTTI and how the framework could be used to produce more-informative readiness metrics. In the aggregate, the following sections provide a review of the state of OTTI.

Competency Model

Competency models *define the skills and capabilities that individuals, units, and the joint force must possess to successfully perform missions.*[12] By identifying the components that underlie successful performance, competency models inform the creation of training scenarios that exercise those components. Additionally, by identifying sources of evidence for demonstrating mastery, competency models may enable personalized proficiency-based training in the place of current one-size-fits-all approaches.

The diversity of competencies increases as the focus expands from individual to collective and joint skills. Basic pilot skills involve narrow perceptual, motor, and cognitive abilities. For example, "Fly at low level—Able to tactically maneuver aircraft at low level while using the terrain and prioritize terrain clearance tasks."[13] Complex individual skills require selecting and integrating multiple knowledge, skills, and supporting competencies. For example, "Prioritizes—Makes decisions about the relative importance of competing demands." Finally, collective and joint skills encompass *individual task behaviors* along with *collective behaviors*, such as communicating, mutual performance monitoring, and maintaining shared situational awareness.[14]

Current Frontier: Individual Skills

The competency model underlying current training practices is implicit in the RTM and other Air Force training publications, such as the AFI 11-2 Volume 1 series for each weapon platform. The RTM lists training requirements for primary and secondary mission areas, along with flight

---

[12] Charles M. Colegrove and George M. Alliger, "Mission Essential Competencies: Defining Combat Mission Readiness in a Novel Way," presented at the NATO RTO Studies, Analysis and Simulation (SAS) Panel Symposium, Brussels, April 2002.

[13] George M. Alliger, Rebecca Beard, Winston Bennett, Jr., Steven Symons, and Charles Colegrove, "A Psychometric Examination of Mission Essential Competency (MEC) Measures Used in Air Force Distributed Mission Operations Training Needs Analysis," *Military Psychology*, Vol. 25, No. 3, 2013.

[14] Heikki Mansikka, Kai Virtanen, Don Harris, and Matti Jalava, "Measurement of Team Performance in Air Combat–Have We Been Underperforming?" *Theoretical Issues in Ergonomics Science*, Vol. 22, No. 3, 2021.

events that pilots must complete on a recurring basis. Other Air Force publications document tactical flying skills and standards and evaluation criteria for each aircraft and weapon system.[15] Collectively, these publications and the criteria set forth in them form a basis for defining job performance in terms of primary and secondary mission areas, for identifying tactical flying skills that underlie effective job performance, and for evaluating job performance relative to standards and evaluation criteria. The collection of Air Force publications does not, however, directly map tactical skills, standards, and evaluation criteria to missions. In addition, the publications do not include all tactical skills or many collective and joint skills that underlie collective and joint performance.

The MEC model provides a complementary framework to directly explicate competencies, knowledge, and skills to allow pilots, units, and joint forces to successfully complete missions in adverse conditions and in nonpermissive environments.[16] MECs specify higher-order individual, collective, or joint competencies, such as the competencies required to engage in air-to-air combat.[17] The MEC framework further specifies supporting competencies, knowledge and skills, and the experiences to engender each.

The MEC process has been applied to numerous platforms, mostly in Air Combat Command, and provides mature competency models for basic and complex individual skills.[18] In addition, the MEC process identifies supporting competencies that contribute to collective behavior, such as communication and coordination, leadership, and teamwork, although these describe the behavior of an individual within the unit rather than the unit itself. Finally, the MEC framework has been applied separately to other Air Force functions, such as command and control; to Navy and Army missions; and to coalition forces. Notwithstanding this generality, MEC development efforts have not primarily focused on joint and coalition missions. For these reasons, the frontier for competency models is currently limited to initial skills and individual task behaviors, and work is underway to specify models for collective and unit behavior.

Technology Horizon and Barriers to Progress

The barriers preventing the Air Force from moving from individual competency models to collective and joint models have primarily to do with a lack of agreement on which models are appropriate for the latter two skill categories. The skills and competencies that aircrew members need to successfully execute their respective tasks in individual training environments and for which they can individually demonstrate proficiency are necessary but not sufficient for the individuals to form a proficient team. In addition to the difficulties in making the jump from

---

[15] For example, the Air Force Tactics, Techniques, and Procedures (AFTTP) 3-3 series discusses tactical flying skills, and AFI 11-2 MDS Volume 2 series presents standards and evaluation criteria.

[16] Alliger et al., 2013.

[17] Alliger et al., 2013.

[18] Note that MECs are currently supported by research and development funds, not operations funds.

individual task performance to collective performance, the flow of pilots into and out of units makes evaluating the performance of stable teams over time even more difficult. The competencies of an individual pilot may translate into better or worse overall collective performance, depending upon the other members of a team.[19]

Although it may be reasonable to assume that individually proficient aircrew members would be *more likely* to form a proficient team than less-proficient individuals, the Air Force lacks models to identify which collective factors—such as communication behaviors or other emergent behaviors representing collective effectiveness—could and should be measured to evaluate collective performance. Such measures are needed to validate implicit models of how individual competencies translate into collective competencies and to build readiness metrics that address collective proficiency.

The same barrier exists for modeling joint competencies, for which issues of measuring and validating proficiency are affected by higher-order complexities. As it readies itself to implement the Joint All-Domain Command and Control (JADC2) concept, the Air Force will need to modify or develop new competency models.[20] JADC2 may entail new interactions between warfighters across domains, new and fluid operational authorities, and more distributed teaming. New models will need to address such questions as the following:

- What does it look like for an individual to be proficient at command-and-control functions in a distributed team?
- What competencies are required to effectively receive and hand off tasking authority in stressful situations?

Given that JADC2 has implications from the strategic to the tactical level of warfighting, its implementation suggests that new competency models may be required across all levels.

Infrastructure

Training infrastructure allows the delivery of live and simulated training events whose content is informed by competency models.[21] Infrastructure needs vary as the training focus expands from individual to collective and joint skills. For example, pilots may use commercial-

---

[19] Erik Gonzalez-Mulé, Bethany S. Cockburn, Brian W. McCormick, and Peng Zhao. Team Tenure and Team Performance: A Meta-Analysis and Process Model," *Personnel Psychology*, Vol. 73, No. 1, Spring 2020.

[20] For a summary of JADC2, see John R. Hoehn, "Joint All-Domain Command and Control (JADC2)," Washington, D.C.: Congressional Research Service, IF11493, March 18, 2021.

[21] While simulation capabilities (see Figure 1) may be considered part of training infrastructure broadly conceived, we separate these out because of the directness of their role in representing the tasks being trained. Additionally, the challenges for generating simulated content and providing the infrastructure used to combine simulated content in a common environment differ substantially in character, are driven by distinct factors, and are addressed by separate technological disciplines.

off-the-shelf augmented and virtual reality systems to rehearse basic skills.[22] Complex individual skills may require simulators that exercise all the skills needed to perform a mission. Simulators must also be physically, functionally, and psychologically representative for skills developed during training to transfer to the operational environment.[23] Collective training requires networked simulators to connect two or more training systems in a shared environment. Finally, collective and joint training may require distributed mission infrastructure to allow geographically dispersed units to train together.[24] Distributed infrastructure presents challenges in terms of linking training systems developed with different standards in a shared environment, supporting multiple security domains or enclaves, and managing network performance to allow systems to interact in real time.

Current Frontier: Collective Skills

Networked simulators are available for unit training but are distributed unevenly across the Air Force, and a number of policy, technical, and resource barriers prevent more widespread, frequent use.[25] Relevant aspects of OTTI depend on whether the training is at the national and combatant command levels (tier 1), joint task force training (tier 2), functional and service component headquarters training (tier 3), or individual organizational training (tier 4).[26]

Fewer networked resources are typically required for tier 1 training. For example, theater-level training is sufficiently broad in scale that highly aggregated simulations using fewer systems can achieve training goals.[27] In contrast, the technical capabilities required to network simulators are greater across tiers 2 through 4 because of the higher degree of interactions occurring between simulation platforms. For example, the number of participants, types of simulation platforms, and scope of data exchange requirements increase substantially as training incorporates more individual and service-level training goals. When these systems are geographically distributed, network authorizations are required for exercise participation, and system interoperation becomes more sensitive to network latency.

---

[22] Barbara Barrett and David L. Goldfein, "United States Air Force Posture Statement Fiscal Year 2021," presentation to the Senate Armed Services Committee, 2nd Sess., 116th Cong., Washington, D.C., March 3, 2020.

[23] Cliff Noble, "The Relationship Between Fidelity and Learning in Aviation Training and Assessment," *Journal of Air Transportation*, Vol. 7, No. 3, 2002.

[24] Chapman and Colegrove, 2013.

[25] John A. Ausink, William W. Taylor, James H. Bigelow, and Kevin Brancato, *Investment Strategies for Improving Fifth-Generation Fighter Training*, Santa Monica, Calif.: RAND Corporation, TR-871-AF, 2011.

[26] Chairman of the Joint Chiefs of Staff Instruction 3500.01J, *Joint Training Manual for the Armed Forces of the United States*, Washington, D.C.: Office of the Chairman of the Joint Chiefs of Staff, January 13, 2020, pp. B-5 and B-6.

[27] Theater-level training is primarily oriented toward command staff of a geographic combatant command, where command-level decisions concern large-scale forces and are of critical significance. Training at the level of individual services emphasizes detailed operational and tactical training that implements commander decisions.

The extent to which individual simulators are and can be networked depends on such factors as differences stemming from simulator-specific capabilities to support data and effects in the operating environment, information security requirements, differences in communication and data standards used for interoperation, and the general nature of the training interactions required for a given training tier (e.g., frequency, level of aggregation). High-fidelity, real-time communication (visual or audio) across multiple platforms is computationally challenging and difficult to implement, but the network must support both this and the accurate representation of effects. Each service maintains integration tools that enable joint interoperability, and the Joint Staff has defined federation methods for cross-service or collective training.[28]

Training capabilities for mission-specific tasks increasingly rely on infrastructure to support complex simulation capabilities, such as managing environmental data, introducing synthetic elements to training (e.g., virtual and augmented reality), flexibly and rapidly preparing scenarios and threats, and supporting multiple levels of security for a given event. These issues are also a concern for Air Force training on physical ranges, such as in a series of flag exercises dedicated to testing capability integration.[29] For these reasons, we place the frontier for infrastructure at the level of networked simulations.

Technology Horizon and Barriers to Progress

The barriers to building the infrastructure for networked training for collective and joint forces in high-complexity environments are nontrivial to resolve technologically and related to organizational and policy challenges. These include the following factors:

- **Network latency.** To accommodate large-scale distributed training, networks need to have sufficiently low latency to provide realistic training. Barriers to achieving low-latency networks include inefficient translation of data exchanged between simulation platforms, distance between networked systems, and data processing speed.
- **Bandwidth.** Training in highly complex environments requires sufficient network bandwidth to accommodate the simulation of many entities and environmental and other effects. For collective and joint training, networks must be able to accommodate many human-in-the-loop simulators or live entities connected to simulated environments.

---

[28] See for example, Joint Staff Directorate for Joint Force Development's Joint Training Synthetic Environment/Joint Training Tools and the Joint Live, Virtual, Constructive Federation (Terrence E. Culton, David W. Parkes, and J. Thomas Walrond, *Future Construct/Architecture for Modeling and Simulation Support to Joint and Collective Training Across the Continuum of Military Operations*, Brussels: North Atlantic Treaty Organization, STO-MP-MSG-143, undated). Broadly, individual simulators are integrated using standardized protocols (e.g., Distributed Interactive Simulation, High-Level Architecture) or directly through vendor-specific frameworks. Existing tools require substantial time and resources to address integration needs and perform testing prior to use in large-scale training exercises. Current standards for interoperation require nuanced interpretation of standards that complicates the ability to integrate systems. Bold Quest provides an example of the scale and character of the interoperability required for training (see, e.g., Jim Garamone, "Bold Quest Event Builds Interoperable Fires for Tomorrow," press release, U.S. Department of Defense website, November 19, 2020).

[29] Christine Saunders and Savanah Bray, "Orange Flag, Black Flag Collaborate to Accelerate Change," press release, U.S. Air Force website, March 10, 2021.

Barriers include the number of participants on a network, the volume of data exchanged between simulation platforms for real-time processing, and the ability of a host facility to upgrade its network infrastructure.

- **Security concerns.** Receiving authority to operate simulators for one platform—much less from other services and other domains—on training networks besides their own is time and resource intensive. Individual training units and offices are reticent to accept the perceived risks of connecting non-native devices to training networks. A number of barriers to widespread distributed training involve risk, including how difficult it is to obtain the authority to operate, differences in perceived security impact when systems are connected, and a lack of policy direction to accept risk or to ease the authority to operate process.[30]

- **Common data standards and architectures.** The lack of shared data standards and architectures within the Air Force, across the services, and across other mission partners is a significant impediment to distributed training.[31] Barriers include the lack of effective coordinating bodies—both service and joint—to ensure common standards and architectures and the absence of genuine joint interoperability requirements that would be enforced in the acquisition process for training systems.[32]

These barriers to providing distributed training at scale also limit the quality of readiness metrics that the Air Force can implement at scale across the range of reporting units. Until the Air Force can provide units with the infrastructure to train in—and report on performance in—complex environments with many Red and Blue actors, there will be limited opportunity for units to provide readiness metrics that speak to performance in such environments.

Simulation Capabilities

Training infrastructure enables the delivery of simulated content using a range of simulation capabilities for both live and virtual training. Training and readiness assessment requires a sufficiently faithful representation of combat environments, including the number and composition of Red and Blue Forces, behavioral models of Red and Blue force TTPs, and natural (e.g., weather) and manmade physical conditions (e.g., electromagnetic spectrum effects). The

---

[30] Although DoD cybersecurity policy (DoD Instruction 8510.01) specifies the application of the Risk Management Framework. And although enclosure 5 of the instruction provides a common *process* for establishing reciprocity, standard *guidance* on how risks are prioritized or how the potential security impact of risks is interpreted is not provided. In practice, this contributes to a time and resource intensive process for establishing authorities to operate and to mitigate risks. See DoD Instruction 8510.01, *Risk Management Framework (RMF) for DoD Information Technology (IT)*, Change 3, Washington, D.C.: U.S. Department of Defense, December 29, 2020.

[31] Emilie A. Reitz and Kevin Seavey, "Making Joint and Multinational Simulation Interoperability a Reality," presented at the Interservice/Industry Training, Simulation, and Education Conference, November–December 2018.

[32] Timothy Marler, Matthew W. Lewis, Mark Toukan, Ryan Haberman, Ajay Kochhar, Bryce Downing, Graham Andrews, and Rick Eden, *Supporting Joint Warfighter Readiness: Opportunities and Incentives for Interservice and Intraservice Coordination with Training-Simulator Acquisition and Use*, Santa Monica, Calif.: RAND Corporation, RR-A159-1, 2021.

simulation capabilities needed change as the focus expands from individual to collective and joint skills.

For example, basic skills can be rehearsed in part-task trainers that abstract many details of the combat environment. Complex individual skills require simulators that represent the adverse conditions and nonpermissive environments under which missions must be completed. Because more actors are involved, collective skills may further require constructive (i.e., computer generated and controlled) agents to simulate adversary and friendly forces. Finally, joint skills place even greater emphasis on simulating friendly and adversary behaviors and effects for forces that operate outside the air domain. As with infrastructure, this presents additional challenges in terms of integrating simulation capabilities across joint systems.

Current Frontier: Individual Skills

Although Air Force training systems can simulate a high degree of environmental complexity (weather effects, kinetic effects, electronic warfare, and others) and induce physically realistic motion in simulators,[33] virtual and augmented-reality systems are being explored to deliver training that is not possible on physical training ranges.[34] Virtual and augmented-reality systems are expected to reduce the cost of training, increase training efficiency, and enable training of a broader range of missions against emerging threats. For example, the simulation of aggressor (adversary) tactics currently relies mostly on live platforms being operated as Red air. Experimental simulation capabilities seek to simulate aggressor capabilities and tactics at distances exceeding the capacity of physical ranges,[35] for nonlocal terrain and environmental effects, and at a scale beyond the Air Force's Red air capacities.

Although aggressor models are in development for advanced training, they are not widely available today and confront open problems that prevent widespread adoption. Simulation capabilities that require extensive networking and real-time processing of large volumes of data are limited by training site infrastructure.[36] Training networks for live- or simulator-based training face challenges with capacity and real-time data processing. These challenges limit the

---

[33] Simulation fidelity is frequently highlighted as a key training need. Fidelity is the degree to which a simulation represents the real operational platform or environment. However, fidelity is also important to consider for its actual contribution to training proficiency. See Susan G. Straus, Matthew W. Lewis, Kathryn Connor, Rick Eden, Matthew E. Boyer, Tim Marler, Christopher M. Carson, Geoffrey E. Grimm, and Heather Smigowski, *Collective Simulation-Based Training in the U.S. Army: User Interface Fidelity, Costs, and Training Effectiveness*, Santa Monica, Calif.: RAND Corporation, RR-2250-A, 2019.

[34] Valerie Insinna, "US Air Force's T-38 Trainer Could Soon Dogfight with Augmented Reality Adversaries," Defense News website, March 19, 2021; Mandy Mayfield, "Virtual, Augmented Reality Tech Transforming Training," National Training & Simulation Association, February 17, 2021.

[35] Defense Advanced Research Projects Agency, "AlphaDogFight Trials Go Virtual for Final Event," press release, U.S. Department of Defense website, August 7, 2020.

[36] Simulation capabilities depend on the technical infrastructure made available by their host facilities to support requirements for data management, networking, and security (e.g., multiple levels of security for cross-platform and joint training).

widespread use of aggressor models and the ability to provide distributed mission training broadly across the Air Force. For these reasons, we place the frontier for simulation capabilities at the level individual skills.

This frontier tracks with current limitations in readiness metrics. Units use simulation capabilities to meet RAP requirements and to inform commanders' subjective assessments through their use in larger-scale exercises; however, simulation (and related infrastructural) capabilities are not currently adequate for units across the Air Force to use them to build metrics on collective and joint performance in common simulated environments.

Technology Horizon and Barriers to Progress

The barriers to simulating a diverse range of Red and Blue models to train units in shared, high-complexity environments are, as with infrastructure, primarily technological, policy, and organizational. Prominent barriers include the following:

- **Common environmental and terrain data.** Participants in a simulation should all see the same thing; for example, a tank should not be on the ground in one simulation and half-submerged in another. There are a number of organizational and policy barriers to the adoption of common environmental and terrain data, including coordinating bodies and policy imperatives.[37] The time and resource costs of ensuring that separate simulators are using common databases on an event-by-event basis can be prohibitive and could serve to limit the Air Force's ability to engage in collective, distributed training.
- **Common conceptual model.** Training in shared simulation environments is enabled by common conceptual models, which are foundational models that describe the elements of a simulation and how they interact.[38] For example, one simulation may model water as a barrier to movement while another may not, or simulations may not share the same criteria for what constitutes a fatal hit on a particular target. Lack of shared conceptual models can impede interoperability and reduce training fidelity and effectiveness. The barriers are similar to those for common databases noted earlier.
- **Adequate threat models.** The ability to provide threat models is impeded by an inherent degree of uncertainty about adversaries' capabilities and tactics and the technical limitations to simulating adversaries discussed earlier.

Data Capture

The Air Force's current and planned simulation capabilities hold the promise of training units in highly complex environments, but the ability to harness these capabilities for performance

---

[37] Although the Air Force manages to synchronize environmental and terrain data in practice, the resource intensiveness negatively affects the number of and ease with which training events are delivered and the range of different platforms that may be integrated.

[38] On conceptual modeling for military simulations, see Research and Technology Organisation, North Atlantic Treaty Organisation, *Conceptual Modeling (CM) for Military Modeling and Simulation (M&S)*, TR-MSG-058, July 2012.

assessment depends on data capture. A range of measures can be recorded during live and simulated flight. These include objective measures, such as aircraft positional and system data; physiological measures, such as pilot heart rate variability; and subjective measures, such as SME ratings of pilot and collective performance.[39] The data needed to assess performance change as the focus shifts from individual to collective and joint skills. Basic skills, such as instrumented flight rules, are linked to observable changes in aircraft position and system state. Aircraft or network data may be captured to assess these skills.

Individuals demonstrate competency in complex skills by selecting and performing contextually appropriate actions. Beyond individual metrics, data about context must be recorded to assess these skills. An added complication is that complex skills depend on pilots' situational awareness and decisionmaking, which cannot be directly observed. Pilot self-reports gathered concurrently or retrospectively during debriefings may provide additional data for assessing these skills. Collective skills encompass task behaviors along with collective behaviors, such as communication. Thus, contextual data, data from multiple aircraft and their behaviors relative to one another, and communication data are needed to assess collective skills. Finally, all these types of data are needed to assess joint performance. However, capturing data from joint and coalition training systems and environments raises additional technical and policy challenges involving latency, bandwidth, security, and interoperability.

Current Frontier: Individual Skills

Capabilities exist to capture data from simulators and simulations but are not routinely used in operational training, but no formal requirements have been set out for capturing and storing data from simulator sorties or from virtual environments and virtual exercises (although they have been used for some live exercises, such as Red Flag).[40] Data captured by simulators can be used to inform task and skill performance assessment. Relevant data about tasks and skills are discussed in briefings and provided as feedback between training sessions. Some efforts are underway to explore data collection from additional sensors (e.g., physiological responses) and in ways that support real-time assessment and training adaptation using machine learning methods.[41] Real-time assessment and training adaptation require the capture and management of large volumes of data to evaluate trainee performance.

Data captured from training systems, such as head-mounted displays or simulators, requires raw data from independent streams to be merged and in a common format with metadata about

---

[39] Jonathan Borgvall, Martin Castor, Staffan Nählinder, Per-Anders Oskarsson, Erland Svensson, *Transfer of Training in Military Aviation. Command and Control Systems*, Kista, Sweden: Swedish Defense Research Agency (FOI), 2007.

[40] Interview with U.S. Air Force officials, January 15, 2021.

[41] Jaclyn Hoke, Christopher Reuter, Thomas Romeas, Maxime Montariol, Thomas Schnell, and Jocelyn Faubert., "Perceptual-Cognitive & Physiological Assessment of Training Effectiveness," presented at the Interservice/Industry Training, Simulation, and Education Conference, November–December 2017.

the source systems. Intermediary software combines raw data to provide contextual information for use in analysis. Although many performance-related data sources exist, personnel involved in training across MAJCOMs did not indicate that any training systems automatically combine and standardize data collection from heterogenous sensors or simulation systems. Information about training performance is presented during post-training briefings, but, at present, the method of collection is not standardized across the Air Force. For all these reasons, we place the OTTI frontier for data capture at the individual skill level.

Technology Horizon and Barriers to Progress

The capture and storage of data from aircrew, simulators, and simulations are not particularly challenging technical problems. Investing in such capabilities is, instead, inhibited by a lack of policy direction—e.g., no formal requirements, funding streams, or leadership structure pushing for the capture of training-system data—combined with a lack of understanding about which data should be captured and retained to contribute to performance assessment across the skill levels. Because data capture and storage are not dictated by policy, the capabilities to capture these data are generally not funded. Capabilities to capture simulation and simulator data exist (e.g., PETS, as discussed earlier) but have not been adopted across the Air Force for these reasons. Apart from issues of policy, knowing what data to capture and store is a prior requirement for investing in this capacity, just as it is a prerequisite for creating and implementing reporting requirements on a new set of readiness metrics that use these data. A requirement to capture simulator data would need to be supported by an underlying model of how that data would permit analysis of aircrew performance of individual and collective tasks. Data capture is thus linked to and limited by the development of competency models that specify what data should be captured.

Data Analysis

Data analysis turns the data produced by simulation capabilities, training devices, and training audiences into information that can be used for a variety of purposes, including readiness assessment. Basic skills are evaluated relative to a standard. The analyses to support these evaluations may involve comparing aircraft data, such as airspeed or turn rate, against predefined thresholds. Complex individual skills further involve selecting contextually appropriate TTPs. Currently, SMEs use domain knowledge and experience to evaluate TTPs manually and subjectively. Automated analysis of TTPs would require specifying how myriad contextual factors relate to the probability of task success given different TTPs and how task success translates to mission outcomes. Analysis of collective behavior is complicated by the even greater complexity of the environment (i.e., multiple individuals interacting with one another). In addition, certain behaviors emerge only from interactions between pilots.

Understanding behaviors that relate to effective teamwork is required to answer such questions as

- Which communication patterns reveal a shared understanding of the threat environment?
- Which communication patterns reveal effective team leadership?
- Which monitoring activities reveal effective backup behaviors?[42]

The same types of analyses are needed to evaluate joint behavior. Automated assessment of joint behavior is further complicated by the still greater complexity of joint and coalition missions and the scarcity of historical data to even identify behaviors that contribute to positive mission outcomes.

Current Frontier: Initial Skills

Numerous automated data analysis capabilities exist, although nearly all have been demonstrated only in research and test environments and not in operational environments. For example, physiological data, such as heart rate and brain activity, can be converted into indices of mental workload, and eye movements can be related to situational awareness.[43] Additionally, psychological batteries, such as the NASA Task Load Index and the Situation Awareness Global Assessment Tool, provide interpretable measures of mental workload and situational awareness. These measures are most meaningful at the levels of basic and complex individual skills.

Additional methods for analyzing performance data exist, but again, few of these methods have been demonstrated in operational environments. Basic skills can be scored using predetermined criteria based on the occurrence and timing of events and aircraft positional data, for example. The MEC process identifies additional measures for assessing knowledge, skills, and competencies (e.g., number, timing, and effect of shots, speed management, and separation).[44] A challenge in the case of complex skills is accounting for the effects of context and other actors' behavior on pilot performance. In addition, certain skills, such as maintaining situational awareness of the battlespace, cannot be directly observed and so must be inferred from supporting evidence. Finally, it may not be possible to catalogue all states of the environment to specify "correct" behaviors and TTPs for complex scenarios in advance.

These challenges are exacerbated in the case of collective and joint performance. A further complication is measuring teamwork. Potential measures of collective constructs, such as shared situational awareness and communication, exist, but the objective assessment of teamwork in the

---

[42] Ulrika Ohlander, Jens Alfredson, Maria Riveiro, and Göran Falkman, "Fighter Pilots' Teamwork: A Descriptive Study," *Ergonomics*, Vol. 62, No. 7, 2019.

[43] Heikki Mansikka, Kai Virtanen, Don Harris, and Petteri Simola, "Fighter Pilots' Heart Rate, Heart Rate Variation and Performance During an Instrument Flight Rules Proficiency Test," *Applied Ergonomics*, Vol. 56, September 2016; Sandro Scielzo, Justin C. Wilson, and Eric C. Larson, "Towards the Development of an Automated, Real-Time, Objective Measure of Situation Awareness for Pilots," presented at the Interservice/Industry Training, Simulation, and Education Conference, November–December 2020.

[44] Schreiber, 2017.

air domain is in its relative infancy. Subjective measures of teamwork also exist but depend on SME ratings of collective performance.[45] Outcome measures, such as accomplishment of mission objectives, may be informative, yet they may not adequately capture variations in collective performance. For all these reasons, we place the OTTI frontier for automated data analysis at the initial skill level.

Technology Horizon and Barriers to Progress

As with competency models, progress on data analysis capabilities is limited by foundational—more so than policy or technical—challenges in understanding how to harness (potentially) available data from simulators and simulations to assess performance across skill levels and to produce associated readiness metrics.

Although the recent push for a proficiency-based aircrew training system has highlighted the need for objective metrics,[46] subjective measurement and SME opinion in which objective measures are inadequate but still have a role to play. However, the Air Force lacks an understanding of where subjective measures are necessary—for reasons of technological maturity or methodological necessity—and where technology is available and appropriate to produce quantitative measures for underlying performance concepts.

The current development horizon falls short of individual skills, where analytic capabilities exist to evaluate individual aircrew TTP selection. Barriers to evaluating TTP selection include the following:

- Identifying the range of factors that are responsible for task and mission success is difficult. Without an adequate accounting of the most important factors that relate TTP selection with outcomes, analytic approaches may not provide reliable answers.
- The lack of large databases hinders the use of quantitative (i.e., machine learning) methods to evaluate how different TTP selections influence task and mission outcomes at different flight junctures and in different scenarios.
- The data sets that might be generated if the U.S. Air Force were to capture data from training events might be insufficient. The utility of many machine-learning methods is limited by the volume of data to train algorithms. Even if the Air Force were to capture all potential data from live and simulated sorties, the stakes of getting the wrong answer place a high bar on gathering sufficiently large amounts of data and on developing sufficient trust in the output of ML algorithms to use the data to evaluate TTPs.[47] Using constructive entities to provide additional data for algorithms to evaluate how TTPs relate

---

[45] Heather M. McIntyre, Ebb Smith, and Mary Goode, "United Kingdom Mission Training Through Distributed Simulation," *Military Psychology*, Vol. 25, No. 3, 2017.

[46] U.S. Government Accountability Office, *Ready Aircrew Program: Air Force Actions to Address Congressionally Mandated Study on Combat Aircrew Proficiency*, Washington, D.C., GAO-20-91, February 2020.

[47] See Sherrill Lingel, Jeff Hagen, Eric Hastings, Mary Lee, Matthew Sargent, Matthew Walsh, Li Ang Zhang, and David Blancett, *Joint All-Domain Command and Control for Modern Warfare: An Analytic Framework for Identifying and Developing Artificial Intelligence Applications*, Santa Monica, Calif.: RAND Corporation, RR-4408/1-AF, 2020, pp. 44–45, on "explainable AI" and human trust in algorithms.

to combat outcomes may be promising, although TTP evaluation should not be based purely on simulated data.[48]

Barriers to data analytic capabilities for collective and joint skills are yet more fundamental. There are no validated methods that can analyze individual metrics and produce information about collective and joint performance;[49] as noted previously, proficient individuals do not necessarily sum to proficient teams. The same issues that hinder analysis of TTP selection apply to the analysis of collective and joint training. In addition to the same environmental factors that shape TTP selection, individual TTP selection itself becomes a relevant factor that influences collective performance.

As noted earlier, existing methods of analyzing aircrew teamwork are underdeveloped, and little validation of existing models has been done. Not only are measures of such things as shared situational awareness cumbersome to capture but including these measures in the analysis of collective and joint performance poses additional requirements: identifying the junctures at which these measures should be taken and specifying and measuring appropriate outcomes. As mentioned earlier, mission-level outcomes are available and informative but may not be well matched to analyzing collective behavior in a manner that permits the adaptive training of specific skills.

---

[48] For details on this approach, see Heikki Mansikka, Kai Virtanen, Don Harris, and Jaakko Salomäki, "Live–Virtual–Constructive Simulation for Testing and Evaluation of Air Combat Tactics, Techniques, and Procedures, Part 2: Demonstration of the Framework," *Journal of Defense Modeling and Simulation*, Vol. 18, No. 4, October 2019.

[49] See Mansikka et al., 2021, for a recent review of existing team performance assessment constructs as applied to air combat. The review highlights that, while aggregate outcomes, such as the ratio of Blue to Red losses, are commonly used as the product of team performance, existing frameworks lack the capability to evaluate team processes, such as the contributions of individual aircrew tactics to the overall output of teams.

5. Overarching Barriers and Concluding Thoughts

As the previous discussion indicated, individual OTTI components are at different levels of maturity and face different barriers to further advancement. As Table 2 illustrates, the horizon for infrastructure, simulation capabilities, and data capture is characterized primarily by challenges in the U.S. Air Force's organizational approach to capability development, policy impediments, or technical issues. In contrast, the challenges to advancing the horizon for competency models and data analysis are more fundamental and reflect a lack of consensus and deeper uncertainty about the best way forward. Thus, the first group represents areas in which the Air Force knows how to harness OTTI components, but various factors limits its ability to do so; the second group represents areas where the Air Force is less sure of how to harness OTTI in the first instance—policy, technology, and organizational barriers aside.

Table 2. Barriers Toward Advancing the Horizons of Operational Test and Training Infrastructure Technologies

Category	Initial Skills	Individual Skills	Collective Skills	Joint Skills
Competency model	Perceptual and motor skills	Task behaviors	Team behaviors	Interteam behaviors
Infrastructure	Trainer	Simulator	Networked simulators	Distributed joint mission training
Simulation capabilities	Low environment complexity	High environment complexity	Red air models	Diverse Blue and Red system models
Data capture and storage	Few performance variables	Many performance variables	Scenario and Red air	Non–Air Force systems
Data analysis	Performance relative to standard	TTP selection	Team behavior (communication)	Interteam behaviors

NOTES: Green shading indicates that the limitations are related to technical, organization, or policy issues. Blue shading indicates that the limitations are more fundamental (lack of consensus, foundational knowledge). The black line indicates current horizon boundaries.

Table 2 presents a framework for identifying how different investments in OTTI advance readiness assessment capabilities across levels, from initial skills up to the joint force. The framework can be used to target and coordinate investments and research efforts, and to track the overall state of OTTI across time. The vertical boundaries mark the current horizon (i.e., the highest levels of capabilities currently achieved and routinely used). The framework shows that the current horizon would permit automated measurement of pilot abilities in the initial skills domain, but the difficult problem of evaluating behaviors in a complex environment, the way a pilot examiner would in mission qualification training, is a barrier to improving on RAP with better indicators of a pilot's T-level and contribution to the ability of a unit to execute its mission-essential tasks. If research efforts overcome that barrier, the next frontier would involve similar but, more complex, research and technological challenges to leveraging individual OTTI components to produce collective metrics. By monitoring the maturity levels of each row in the framework, readiness decisionmakers can understand which areas limit the development of better metrics as new technologies and research breakthroughs move the horizon to the right.

The diverse nature of challenges within and across OTTI components underlines the importance of coordinating development strategies. Progress along each dimension of OTTI is not independent of the others; barriers are interrelated across components; and progress on one dimension often hinges on progress along the others. For example, improvements to simulation capability would have the benefit of delivering more-authentic training, but that goes only so far in providing readiness information for commanders and senior leaders.[1] More-authentic training scenarios and environments need to be developed alongside accurate models of collective performance measurement, underlying competency models, and analytic capabilities that are matched to the data collected. More-authentic training offers clear benefits, but they are unlikely to be maximized absent underlying competency models.

Competency models for collective training should also serve as a guide for developments in data capture capabilities and should help address how much data and the level of granularity or abstraction needed to provide insight on collective and joint readiness to fight in complex environments. A fire hose of data from simulators and simulations will not be informative—even with the use of sophisticated artificial intelligence and machine-learning techniques—without accompanying competency models to guide data generation and analysis. The frontier for data analysis falls short of even individual skills and the ability to assess TTP selection; this suggests that the Air Force will continue to rely on subjective judgments when assessing individual TTP, even as the other elements are in place to automate and track TTP performance.

The analysis here also suggests that investing in individual performance measurement will have limited returns absent a set of analytic tools to aggregate metrics. Because combat power is, at some point, delivered jointly, increasing insight into individual performance will not yield additional insights on the ability of units to perform complex, interdependent combat functions. Competency models and data analytic techniques appropriate to the models are thus crucial for maximizing the utility of networked training infrastructure and the data generated in distributed training events: enabling readiness assessment at the service or total force level.

Competency models and data analytic techniques are not themselves sufficient for the assessment of collective and joint skills without training systems that capture and store data in an interoperable fashion. The acquisition of OTTI systems frequently uses external vendors and includes proprietary technologies and information and vendor-specific business processes.

---

[1] Related to training authenticity, too strong a focus on simulator fidelity may detract from a more complete readiness ecosystem. Not all tasks and skills require high-fidelity simulation. Given that fidelity can come at a high cost, including limiting the number of scenarios that it is able to train, the U.S. Air Force may not want to invest in any more fidelity than is necessary.

Incentives may not always align for industry to provide the most interoperable training systems to meet the Air Force's goals of large-scale, distributed training.[2] Similarly, data collected by simulators may not be readily accessible without contractual agreements that define acceptable use of the system's data. These issues can impede the modification of a training system to maintain currency with changes to the platform being simulated, the ability to network systems, or the extraction of data for use in performance assessment.

An overarching barrier to progress along multiple dimensions of OTTI is that training units are incentivized to use their limited resources to pursue unit-specific training goals, given that they are responsible for meeting unit-specific training requirements.[3] This contributes to development of OTTI capabilities that are service specific and a general lack of coordination throughout the training enterprise. Capabilities to replicate training in complex environments alongside joint (much less coalition) actors are therefore scarce relative to their importance. Moving to a world in which the Air Force can measure readiness at more-aggregate levels requires such capabilities.

Table 2 highlights the significant work remaining for the Air Force to produce a training system that can speak—across mission areas and weapon systems—to the readiness of units to perform complex tasks in stressful environments. Note that only one row has a frontier that crosses into the collective skills column, with none crossing into the joint skills column. To the extent that OTTI continues to have these limitations, senior leaders and commanders will be constrained in their ability to assess not just collective performance but, especially joint performance. As the military moves toward operating concepts that involve novel cross-domain interactions—e.g., as envisioned under JADC2—the ability to practice and assess the tactics and skills to execute such a vision will depend on advancing the frontier along all dimensions of OTTI.

In summary, we found the following:

- The frontier of OTTI development generally falls short of providing a capability to assess skills beyond the individual level.
- Progress in certain areas of OTTI—competency models and data analysis—is limited by a lack of foundational knowledge and a lack of consensus; others are held back by organizational, policy, and technical challenges.
- Beyond fundamental research challenges, the lack of incentives to develop interoperable and standardized training systems and the parochial focus of training investments within the Air Force and across the DoD hinder progress along multiple dimensions of OTTI.
- Progress along different dimensions of OTTI is highly interdependent; for example, the development of data storage and analytic capabilities for assessing readiness in highly complex environments depends on identification of competency models and associated performance criteria.

---

[2] On simulation interoperability and incentives, see Marler et al., 2021.

[3] Marler et al., 2021.

- More-authentic training at the individual level alone is unlikely to provide an accurate capability to assess collective and joint readiness; more research is required to identify competencies, metrics, and data analytic capabilities to assess performance and readiness at higher levels of aggregation.

Abbreviations

AFI	Air Force Instruction
CRS	Chairman's Readiness System
DMO	distributed mission operations
DoD	U.S. Department of Defense
JADC2	Joint All-Domain Command and Control
JADO	Joint All-Domain Operations
MAJCOM	major command
MDS	mission design series
MEC	Mission Essential Competency
NASA	National Aeronautics and Space Administration
OTTI	operational test and training infrastructure
PETS	Performance Evaluation Tracking System
RAP	Ready Aircrew Program
RTM	RAP tasking memorandum
SME	subject-matter expert
TTP	tactics, techniques, and procedures

References

AFI—*See* Air Force Instruction.

Air Force Doctrine Note 1-20, "USAF Role in Joint All-Domain Operations," March 5, 2020.

Air Force Instruction 11-200, *Aircrew Training, Standardization/Evaluation, and General Operations Structure*, Washington, D.C.: U.S. Department of the Air Force, September 21, 2018.

Air Force Instruction 10-201, *Force Readiness Reporting*, Washington, D.C.: U.S. Department of the Air Force, December 22, 2020.

Air Force Instruction 10-204, *Air Force Service Exercise Program and Support to Joint and National Exercise Program*, Washington, D.C.: U.S. Department of the Air Force, April 12, 2019.

Air Force Instruction 16-1007, *Management of Air Force Operational Training Systems*," Washington, D.C.: U.S. Department of the Air Force, October 1, 2019.

Alliger, George M., Rebecca Beard, Winston Bennett, Jr., Steven Symons, and Charles Colegrove, "A Psychometric Examination of Mission Essential Competency (MEC) Measures Used in Air Force Distributed Mission Operations Training Needs Analysis," *Military Psychology*, Vol. 25, No. 3, 2013, pp. 218–233.

Ausink, John A., William W. Taylor, James H. Bigelow, and Kevin Brancato, *Investment Strategies for Improving Fifth-Generation Fighter Training*, Santa Monica, Calif.: RAND Corporation, TR-871-AF, 2011. As of December 15, 2021: https://www.rand.org/pubs/technical_reports/TR871.html

Barrett, Barbara, and David L. Goldfein, "United States Air Force Posture Statement Fiscal Year 2021," presentation to the Senate Armed Services Committee, 2nd Sess., 116th Cong., Washington, D.C., March 3, 2020.

Beaubien, Jeffrey M., Michael Tolland, and Jared Freeman, "Performance Measurement Applications and Associated Data Requirements for LVC Training," presented at the Interservice/Industry Training, Simulation, and Education Conference, November–December 2020.

Betts, Richard K., *Military Readiness: Concepts, Choices, Consequences*, Washington, D.C.: Brookings Institution, 1995.

Borgvall, Jonathan, Martin Castor, Staffan Nählinder, Per-Anders Oskarsson, Erland Svensson, *Transfer of Training in Military Aviation. Command and Control Systems*, Kista, Sweden: Swedish Defense Research Agency (FOI), 2007.

Brown, Charles Q., Jr., and David H. Berger, "Redefine Readiness or Lose," War on the Rocks website, March 15, 2021.

Carpenter, Paul E., *Getting More Bang for the Buck: Incentivizing Aircrew Continuation Training*, Maxwell Air Force Base, Ala.: Air University, December 2016.

Chapman, Anne W., *The Army's Training Revolution, 1973–1990: An Overview*, Fort Monroe, Va.: U.S. Army Training and Doctrine Command, 1991.

Chapman, Robert, and Charles Colegrove, "Transforming Operational Training in the Combat Air Forces," *Military Psychology*, Vol. 25, No. 3, 2013, pp. 177–190.

Chairman of the Joint Chiefs of Staff Instruction 3500.01J, *Joint Training Manual for the Armed Forces of the United States*, Washington, D.C.: Joint Chiefs of Staff, January 13, 2020, pp. B-5 and B-6.

Chairman of the Joint Chiefs of Staff Guide 3401D, *CJCS Guide to the Chairman's Readiness System*, Washington, D.C., November 15, 2010.

Colegrove, Charles M., and George M. Alliger, "Mission Essential Competencies: Defining Combat Mission Readiness in a Novel Way," presented at the NATO RTO Studies, Analysis and Simulation (SAS) Panel Symposium, Brussels, April 2002.

Culton, Terrence E., David W. Parkes, and J. Thomas Walrond, *Future Construct/Architecture for Modeling and Simulation Support to Joint and Collective Training Across the Continuum of Military Operations*, Brussels: North Atlantic Treaty Organization, STO-MP-MSG-143, undated. As of January 11, 2021:
https://www.sto.nato.int/publications/STO%20Meeting%20Proceedings/
STO-MP-MSG-143/MP-MSG-143-20.pdf

Defense Advanced Research Projects Agency, "AlphaDogFight Trials Go Virtual for Final Event," press release, Defense Advanced Research Projects Agency website, August 7, 2020.

Department of Defense Instruction 8510.01, *Risk Management Framework (RMF) for DoD Information Technology (IT)*, Change 3, Washington, D.C.: U.S. Department of Defense, December 29, 2020.

Garamone, Jim, "Bold Quest Event Builds Interoperable Fires for Tomorrow," press release, U.S. Department of Defense website, November 19, 2020.

Gonzalez-Mulé, Erik, Bethany S. Cockburn, Brian W. McCormick, and Peng Zhao, "Team Tenure and Team Performance: A Meta-Analysis and Process Model," *Personnel Psychology*, Vol. 73, No. 1, Spring 2020, pp. 151–198.

Greenwood, Thomas C., Terry Heuring, and Alec Wahlman, "The 'Next Training Revolution': Readying the Joint Force for Great Power Competition and Conflict," *Joint Force Quarterly*, Vol. 100, 1st Quarter 2021, pp. 26–34.

Harrison, Todd, "How Many Flying Hours Does It Take To Kill a Terrorist?" Defense One website, November 17, 2014.

Hecker, James B., "Joint All-Domain Operations," *Air & Space Power Journal*, Summer 2021, pp. 2–4.

Hoehn, John R., "Joint All-Domain Command and Control (JADC2)," Washington, D.C.: Congressional Research Service, IF11493, March 18, 2021.

Hoke, Jaclyn, Christopher Reuter, Thomas Romeas, Maxime Montariol, Thomas Schnell, and Jocelyn Faubert, "Perceptual-Cognitive & Physiological Assessment of Training Effectiveness," presented at the Interservice/Industry Training, Simulation, and Education Conference, November–December 2017.

Insinna, Valerie, "US Air Force's T-38 Trainer Could Soon Dogfight with Augmented Reality Adversaries," Defense News website, March 19, 2021.

Lingel, Sherrill, Jeff Hagen, Eric Hastings, Mary Lee, Matthew Sargent, Matthew Walsh, Li Ang Zhang, and David Blancett, *Joint All-Domain Command and Control for Modern Warfare: An Analytic Framework for Identifying and Developing Artificial Intelligence Applications*, Santa Monica, Calif.: RAND Corporation, RR-4408/1-AF, 2020. As of January 12, 2022: https://www.rand.org/pubs/research_reports/RR4408z1.html

Link, Edwin A., Jr., Combination Training Device for Student Aviators and Entertainment Apparatus, U.S. Patent 1,825,462, filed March 12, 1930.

Page, Ray L., "Brief History of Flight Simulation," presented at SimTecT 2000, Sydney, 2000, pp. 11–17.

Mansikka, Heikki, Kai Virtanen, Don Harris, and Matti Jalava, "Measurement of Team Performance in Air Combat–Have We Been Underperforming?" *Theoretical Issues in Ergonomics Science*, Vol. 22, No. 3, 2021, pp. 338–359.

Mansikka, Heikki, Kai Virtanen, Don Harris, and Jaakko Salomäki, "Live–Virtual–Constructive Simulation for Testing and Evaluation of Air Combat Tactics, Techniques, and Procedures, Part 2: Demonstration of the Framework," *Journal of Defense Modeling and Simulation*, Vol. 18, No. 4, October 2019, pp. 295–308.

Mansikka, Heikki, Kai Virtanen, Don Harris, and Petteri Simola, "Fighter Pilots' Heart Rate, Heart Rate Variation and Performance During an Instrument Flight Rules Proficiency Test," *Applied Ergonomics*, Vol. 56, September 2016, pp. 213–219.

Marler, Timothy, Matthew W. Lewis, Mark Toukan, Ryan Haberman, Ajay Kochhar, Bryce Downing, Graham Andrews, and Rick Eden, *Supporting Joint Warfighter Readiness: Opportunities and Incentives for Interservice and Intraservice Coordination with Training-Simulator Acquisition and Use*, Santa Monica, Calif.: RAND Corporation, RR-A159-1, 2021. As of December 15, 2021:
https://www.rand.org/pubs/research_reports/RRA159-1.html

Mayfield, Mandy, "Virtual, Augmented Reality Tech Transforming Training," National Training & Simulation Association, February 17, 2021.

McIntyre, Heather M., Ebb Smith, and Mary Goode, "United Kingdom Mission Training Through Distributed Simulation," *Military Psychology*, Vol. 25, No. 3, 2017, pp. 280–293.

Noble, Cliff. "The Relationship Between Fidelity and Learning in Aviation Training and Assessment," *Journal of Air Transportation*, Vol. 7, No. 3, 2002, pp. 33–54.

Ohlander, Ulrika, Jens Alfredson, Maria Riveiro, and Göran Falkman, "Fighter Pilots' Teamwork: A Descriptive Study," *Ergonomics*, Vol. 62, No. 7, 2019, pp. 880–890.

Research and Technology Organisation, North Atlantic Treaty Organisation, *Conceptual Modeling (CM) for Military Modeling and Simulation (M&S)*, TR-MSG-058, July 2012.

Reitz, Emilie A., and Kevin Seavey, "Making Joint and Multinational Simulation Interoperability a Reality," presented at the Interservice/Industry Training, Simulation, and Education Conference, November–December 2018.

Saunders, Christine, and Savanah Bray, "Orange Flag, Black Flag Collaborate to Accelerate Change," press release, U.S. Air Force website, March 10, 2021.

Scales, Robert H., "The Second Learning Revolution," *Military Review*, Vol. 86, No. 1, 2006, pp. 37–44.

Schreiber, Brian T., "Transforming Training: A Perspective on the Need for and Payoffs from Common Standards," *Military Psychology*, Vol. 25, No. 3, 2017, pp. 294–307.

Schreiber, Brian T., Eric Watz, Antoinette M. Portrey, and Winston Bennett, Jr., "Development of a Distributed Mission Training Automated Performance Tracking System," in *Proceedings of the Behavioral Representations in Modeling and Simulation (BRIMS) Conference*, Scottsdale, Ariz., 2003, pp. 301–317.

Scielzo, Sandro, Justin C. Wilson, and Eric C. Larson, "Towards the Development of an Automated, Real-Time, Objective Measure of Situation Awareness for Pilots," presented at the Interservice/Industry Training, Simulation, and Education Conference, November–December 2020.

Straus, Susan G., Matthew W. Lewis, Kathryn Connor, Rick Eden, Matthew E. Boyer, Tim Marler, Christopher M. Carson, Geoffrey E. Grimm, and Heather Smigowski, *Collective Simulation-Based Training in the U.S. Army: User Interface Fidelity, Costs, and Training Effectiveness*, Santa Monica, Calif.: RAND Corporation, RR-2250-A, 2019. As of December 15, 2021:
https://www.rand.org/pubs/research_reports/RR2250.html

Tucker, Patrick. "An AI Just Beat a Human F-16 Pilot in a Dogfight—Again," Defense One website, August 20, 2020.

U.S. Government Accountability Office, *Ready Aircrew Program: Air Force Actions to Address Congressionally Mandated Study on Combat Aircrew Proficiency*, Washington, D.C., GAO-20-91, February 2020.